Dr. Quantum's
Little Book of
Big Ideas

Dr. Quantum's Little Book of Big Ideas

Where Science Meets Spirit

Fred Alan Wolf

Moment Point Press
Needham, Massachusetts

Moment Point Press, Inc.
PO Box 920287
Needham, MA 02492
www.momentpoint.com

Distribution: Red Wheel Weiser, *www.redwheelweiser.com*

Library of Congress Cataloging-in-Publication Data
Wolf, Fred Alan.
Dr. Quantum's little book of big ideas : where science meets spirit /
Fred Alan Wolf.
p. cm.
ISBN-13: 978-1-930491-07-6 (pbk. : acid-free paper)
ISBN-10: 1-930491-08-5 (pbk. : acid-free paper)
1. Quantum theory—Popular works. 2. Physics—Psychological
aspects—Popular works. I. Title.
QC174.12.W638 2005
530.12--dc22
2005021659

Printed in the United States on recycled acid-free paper
10 9 8 7 6 5 4 3 2 1

A Mini Intro to a Little Book

In the little collection you hold in your hand you will find carefully chosen and arranged excerpts from my books and interviews. I hope you will ponder them, enjoy them, maybe even laugh. They aren't necessarily there to "teach" you anything (although I hope you learn a bit about yourself and the universe) but rather to help you let go of any tight bonds you have created to what you believe is the only reality. You will begin to see that the universe, including you, is far greater than any of us can imagine and that you are a remarkable being—simply because over the last one million or so years of evolution, God has chosen you to appear on the scene!

You may feel that you are not worth very much, or you may feel you are far better than many of your associates. I caution you: Both of these positions are illusions. I hope that by reading this little book you may gain a better perspective on who you really are.

Care to guess?

Asking yourself the deeper questions opens up new ways of being in the world. It brings in a breath of fresh air. It makes life more joyful. The real trick to life is not to be in the know, but to be in the mystery.

I remember a particular day when I was playing in the front hallway of my apartment building. I was barely eight years old. I stood at the top of the stairwell and looked down wondering if I could fly down the nineteen or twenty stairs reaching to the ground floor from our first-floor apartment. Without thinking, I skidded down the stairwell with my feet only barely touching the leading edges of each step. I was on the ground floor in a flash, and I had not slid down the banister, nor had I placed my feet on any of the steps.

When I grew older and remembered what I had done that day, I realized it was impossible. My feet just were not long enough to go from one step edge to the next without my falling flat on my face. Was this just a dream of super powers, or had I actually skidded down those stairs?

quantum

So why is understanding the quantum universe important? Because it excites a feeling of delightful deliciousness — life takes on a new vision. First you find out why it is that you may think the way you think and what conditions have arisen to make you feel that way, and then you can find out how to change them. Quantum physics offers new ways of thinking and new ways of being in the world, which I think improve the various things you do, whether it's taking a walk, talking to somebody, or driving your car. To anything you may basically feel is not enjoyable for you, it can add an aura of joy, because the new metaphors provide fresh ways of thinking and approaching every aspect of your life.

physics

Quantum physics is also important because it contributes to the amazing spectrum of new technologies. I don't think there's a device around that doesn't have a little computer in it somewhere, and that is quantum physics in action. The whole digital age is impossible without quantum physics. And, quantum physics' effect on biology and medicine is enormous. The biggest scientific problem facing our twenty-first-century culture is the relationship of mind to body. With all of the new medicines coming out, and the new insights we're gathering about what constitutes health, quantum physics may just be what we need to really grasp how ancient spiritual views of the body and modern scientific views prove that consciousness can alter reality, and so all illnesses may become as outdated as smallpox is today.

Quantum physics . . .

consists of a well-defined set of rules which work in a universal way. Yet what it predicts about the world is not how the world appears. It predicts, among other things, strange overlaps of reality, parallel realities, and objects being in two or more places at the same time.

provides new metaphors to understand life's experiences.

can change your life!

The movement of life may at times seem chaotic, but it is not. It moves like a gigantic wave and we are caught atop it. If we try to direct the wave, we'll find ourselves in constant battle with it. If we learn merely to survive in the wave's wake, we become victims of it. But we do have another option: We can grasp the rules of the wave's movement and learn to surf it skillfully.

To attempt to become spiritually enlightened without realizing the world of mythology within us is a serious mistake. People who attempt this often find themselves "in battle with the devil" or "in fear of evil."

the power of myth

When that first curtain opened to God's Magic Theater, a great void appeared. And then, according to one myth called Science, the void exploded into the Big Bang. Following another myth called the Bible, in the beginning there was the Word and the Word was with God, and the Word was God. These seemingly very different points of view—these myths called the Big Bang and the Word—appear entirely irreconcilable: one deals with the physical universe of matter and energy and the other with the mental universe of mind and information. But, could these two views actually be saying the same thing? Could it be that in some way what we describe about the universe—how we exploit it to derive meaning from it, how we determine what it is, and what it is doing—establishes the very universe we speak and write about? Does the act of learning something, turning our experiences into meaningful symbols of discourse, create both the physical thing being observed and the laws of order it seems to obey?

The answers to these and other questions will come from a harmonizing of the relationship between the "in here" world of information, meaning, and knowledge with the "out there" world of matter, energy, and existence. This reconcilement is precisely what I mean by a new alchemy.

Perhaps we could move closer to proving the existence of mind by asking: Does matter influence mind?

Well, we can observe matter influencing matter when a dropped egg hits the floor—certainly the floor drastically affects the egg. And when we take a mind-altering substance such as alcohol or a pain-numbing drug, we feel its effect on our ability to be alert or feel pain. So, we tend to believe that matter influences mind. We could say, therefore, that mind and matter exert a force on one another. But since mind cannot be scientifically discovered, this force cannot be found strictly via the fields of physics or physiology or even psychology. No, we need a new field, the field of new alchemy, which incorporates the overlapping insights gained from all three of these fields; which seeks to reveal the secret connection between matter and mind—seeks to reveal the force that we intuitively believe real.

We all seem to know what we mean when we say that an event has occurred. We mean that something happened. We also mean (although we don't usually say so) that someone has observed the event. We take it for granted that events *could* have occurred, even though no one was there to observe them. After all, that proverbial tree we come across, lying on its side in the forest, must have fallen with a great crash of sound as it thumped the ground in its grand descent. Events such as sound-wave production *must* happen regardless of the presence or absence of beings with ears or other organs that can sense these events. Right?

Such a viewpoint is quite natural and is called objectivism. This philosophy states that all reality arises objectively, externally, and independently of the mind. Knowledge of this reality comes from reliably based observations of objects and events—things that happen to these objects. But suppose science proved that objectivism was wrong. Then what? It would mean that all reality is not objective; that reality has to have a subjective quality to it (called mind); and that mind has to affect and possibly even change what we sense as objectively "out there." In brief, it would mean that there is no absolute "out there" unless there is a mind "in here" that perceives it.

The observer effect says that there is no reality until that reality is perceived. This profound insight tells us that we alter every object in the world simply by paying attention to it. In this alteration, both the object of our attention and the mind of the observer change. Because we usually don't pay attention to ourselves in the perception process, our immediate experience usually won't seem to indicate that our actions of perception changed anything. However, if we construct a careful history of our perceptions, they often show us that our way of perceiving indeed changes the course of our personal histories.

Thus the world is really not as it seems. It certainly seems to be "out there" independent of us, independent of the choices we might make. Yet quantum physics destroys that idea. What is "out there" depends on what we choose to look for.

Reality is not just the physical world; it's the relationship of the mind with the physical world that creates the perception of reality. There is no reality without a perception of reality.

Would you be here, exist in physical form, if no one observed you? In a real sense, the answer is no.

Mere observation is enough to alter the history of anything or anyone, even a whole country.

observer effect

By observing, each observer separates into a self and a thing. Often that thing is one's own face, body, or personality/belief structure.

The very notions of heaven being separate from earth, mind separate from body, free-will separate from determinism, life separate from death, and in fact all duality, all opposites, wherein we pose an inside and outside, a boundary line, a nation, an island, a membrane, a distinction—all and more, are not primary facts.

Yet, we unconsciously strive to keep this secret buried inside ourselves. We unwittingly work at maintaining the status quo. We unconsciously choose to live under the illusion that everything is as we see it. This is not only a fundamental truth for you and me, it is the deep secret of the universe's existence: Hide from one's essential self. It is God's great trick; and it only works because we agree to believe the trick. If we can stop believing it for one minute, one second, even one millisecond, and allow our consciousness to become aware that we have stopped, we will see the trick revealed.

At some point in our lives, somehow, somewhere, just for an instant, the unveiling of the great mystery comes to pass. God, the magician, raises the curtain, reveals the trick just slightly, and we glimpse the illusion. But, we don't shout, Wow! No gasps of wonderment fill the theater. Something becomes distinguishable from nothing in a single creative act, but we trick ourselves into not seeing. And so it goes. No applause fills the air. We sit back, watch the show, breathe a sigh of relief, and say unconsciously, "We'll never figure this one out, might as well just accept it."

Distinctions are not real. They are fleeting whispers of an all-pervading, subtle, non-expressive potential reality. The world is not made of separate things. Mind is not separate from matter. And you are not separate from any other being, animal, vegetable, living, dead, or seemingly inanimate matter. The kingdom of heaven and the island of hell lie in you. In you lies everything you have always wanted to know. A vast potential urging itself to arise and become something lies in you. In you, like a coiled serpent waiting to spring forth from your deepest shadows, lies every creative moment that exists, has ever been, and will ever be.

This has been, is, and will always be, the secret of creativity: breaking free of separation.

you-niverse

Within your own mind and body lies a majestic story filled with drama, pathos, humor, intelligence, fantasy, and fact. It is no less than the story of the entire universe, particularly its own creation, transformation, and ultimate purpose. And while most stories require a separated listener and a storyteller, in your story the listener and the storyteller are one. The way in which you go about telling a story to yourself—a story that includes *you*—actually points out that without you there wouldn't be a universe! This story called *you* unfolds into a panorama of life, literally a *you-niverse*. And it involves the sacred transmutation of mind into matter.

Why do we have this essential quality of being that we all feel and can speak of, and resonate with, with a single word called "I"? Everyone seems to have this very definite sense of I-ness, even though I can't necessarily go inside of your body and feel your I-ness. My speculative thought about it is that what I'm feeling as my I-ness is exactly the same as what you feel as your I-ness—that there really isn't any separation there. I-ness is the one quality which, although it appears to divide us, is actually unifying. It's a unity of our common experience.

We walk around with the unconscious ideas that we've been taught, which define us as separate, distinct: "I am this; I am not that. I am good at this; I am not good at that. I am wonderful; I am terrible." These ideas have been ingrained in us since childhood and are re-inforced by others, often family members, who have known us for a long time. But, they don't know us! They only know what they choose to see and think about us. And we don't know them! We only know what we choose to think and see about them.

These are the story lines of our existence. They are not mere wisps of non-inert energy. They are weighty. We embody them, so to speak, and after a long while we carry them with us as if we are walking around with a mass on our shoulders. We have each bought and sold ourselves on certain stories and in our discomfort they bring us a sense of comfort. Any idea or thought that runs counter to our stories, even if it could improve our emotional or financial situation, we quickly push down into a safe, comfortable zone of unthinkability. We have learned, without thinking, how to transform information into matter. In order to break free, in order to have a new experience, a shamanic awakening, a new vision, we have to break free of the illusion that we are separate from anything else—in particular, that we are separate from what we desire. This has been, is, and will always be, the secret of creativity: breaking free of separation. If we don't do that, we don't create.

Matter arises from mind—a vast field of influence commonly envisioned as the Mind of God.

There really is only one Mind, and the thoughts that you're having, while they may seem very personal to you and very self-contained within your own head at the moment, those thoughts are being thought everywhere by everybody at some time in some form. Not necessarily in the same way that you're thinking now, but they're part of a collective consciousness— or of the One Mind.

What is "my self"? Why do I even recognize myself as an individual separate from other individuals? Well, let's look at it this way. If I am a member of a tribe, my concept of self is different than if I am, say, a member of a closed family unit. My behavior, in turn, depends on how I see myself. For example, soldiers fighting in a war view themselves as part of a larger unit; and they behave quite differently toward those outside of their unit, particularly enemies, than they would if they were village people greeting strangers. Thus, how we become aware of the world around us is to a large extent dependent on how we think about our individuality in relation to our environment.

Think about what group or "tribe" you most identify with and how this identification determines your thoughts, beliefs, and behavior. What happens when you change your tribal affinity? A new "you" emerges.

Everyone you see "out there" in life or in your dreams is just a mirror of your own unmasked self. Each is "you," although each wears a different mask/persona, each deals with his or her own karma, seeks his or her own dharma, vows to do, to live, and to play out the role assigned to him or her by the cosmic screen actors' guild.

Our thoughts are representative of the universe's thoughts.

The principle of complementarity says that when you look at something a certain way, that's what you see; when you look at it another way, you see something different.

So, just because you see things one way doesn't mean that's the way they really are. You can see them another way—it may be a very different way from the first way—but it doesn't mean that the second way is the way they really are either! As Walt Whitman said, "Do I contradict myself? Very well then I contradict myself. I am large. I contain multitudes." Each of us is large, containing multitudes! We are not just confined to our bodies.

Only in the great multiplicity of life can we see the One.

principle of complementarity

The principle of complementarity says that the physical universe can never be known independent of an observer's choices of what to observe. Moreover, these choices fall into two distinct, or complementary, sets of observations called "observables." Observation of one observable always precludes the possibility of simultaneous observation of its complement. For example, the observation of the location and the observation of the motion of a subatomic particle form complementary observables. Hence the observation of one renders the other indeterminate or uncertain. So the more objective we are in our observations, the more difficulty we will have in dealing with spirit, and the more likely we will become drawn into the material world. Conversely, as we become more spiritually awakened, the less concern we will feel for our material existence.

By choosing to see the world one way, complementary ways of experiencing become hidden or inaccessible. While these hidden views no longer appear to the mind as objective qualities visible in the world as things, and remembered in the mind as memories, they remain part of the unconscious mind-world as possibilities. Upon a change of choice that brings forward the complementary way of experiencing, these previous hidden views become apparent—"out there"—while the previous qualities vanish in the physical world but remain as the contents of memory. A magician uses this trick to fool us into seeing a thing as it was prior to his sleight of hand.

Our choices—no matter how logical and meaningful they may appear—are among sets of what physicists call "complementary variables." These variables literally have magical and far-reaching consequences, actually altering past, future, and present moments of our lives. Yet because of the illusory normalcy of these observations, we appear to be controlled by them and not the other way around.

Sometimes it's difficult to realize that you always have a choice in everything you do.

In doing quantum physics we had to begin to look at not only a possible way that objects were behaving, but all possible ways an object could behave. For example, if an object goes from A to B, we might, in the normal way of thinking, think of it as following a path—a trajectory, like a straight line or a curve. If you hit a baseball it follows a curved line, if you throw a ball it follows a kind of parabolic arch, as when you throw a football, for example. We can understand those kinds of things.

The quantum way of describing it is that when you throw the ball it follows every possible path you can conceive of to get from A to B—you have to take all those paths into account if you want to know where it will go. And it turns out that you needed all these paths, including imaginary ones that you certainly didn't see, because they helped you explain what you finally did see when you did look.

parallel universes

As fantastic as it may sound, the parallel universe
theory posits that there exists, as if on a different but
parallel layer, another world, a parallel universe, a
duplicate copy, slightly different and yet the same as this
one. And not just two parallel worlds, but three, four,
and even more—no less than an infinite number of
them make up a universe of universes. In each of these
universes, you, I, and all the others who live, have
lived, will live, will have ever lived, are alive.

The thing you call your mind is in itself an infinite number of parallel realities and you are that which looks at all of them together, or separately, depending on whether you focus in, or don't focus in.

We each are composed of large numbers of selves.

Even though we don't experience a flipping back and forth between universes, it may actually be taking place. We might go to sleep and wake up in another universe, but we'd never know because whatever was happening in that universe would seem consistent and logical to us. We wouldn't know we'd made the leap!

Can we awaken in other worlds while we sleep? It just might be that while we're dreaming our minds (not being fully occupied by the world we're in—after all we're dreaming) may be able to perceive these other universes. I'm just speculating, but maybe the key to traveling to another universe is to simplify the mind. The world may just have too many distractions. Perhaps the whole idea is to make these journeys into other universes in our dreams.

Parallel universes have effects one upon the other.
We're in all the universes at once.

Just as a horse becomes overwhelmed by a busy road when its blinders are removed, so would you if you saw that your experience was simultaneously occurring in an infinite number of universal layers where each layer contained a parallel you!

Quantum mechanics is a very strange business. It's stranger than we imagine. It may be stranger than we *can* imagine!

We all believe in the value of hindsight. We've all relived events in our memories, seeing them as we wish they might have happened. Well, imagine that you're able not only to look back in time, but actually to go back in time and make adjustments.

See yourself nudging some event that had dire consequences for you or for another. This nudge needn't take much energy. It could be a mere look on your face witnessed by another, or perhaps a slight change in your voice when speaking to that other. Quantum physics allows us this type of mind-freedom, but it's not quite as simple as I make it sound. In other words, it's possible, in a certain quantum physics sense, to alter the past by changing the way it exists in memory. It isn't that we really go back in time; it's more that we add details to the events that were not sufficiently specified at the time they occurred. For example, we could determine by which means an atom journeys from one place to another by altering how it's perceived at the end of its trip. One choice of that perception would "create" the past history quite differently from a second choice we might have made. In this subtle manner, we create a past from possible counter-factual histories.

If this is true, what is time really all about?

There has never been an adequate definition, a clear metaphor, or even a good physical picture of what time is. The closest we come to observing time is observing what the Buddhists call "being-time." The closest I can come to describing this experience is to use the word "here." Time has nothing to do with movement. Things move, but time stands still. When we say time passes, we mean that we pass. When we say we observe the passing of time, we're doing no such thing—we're observing the passing of our minds, the "movement" of our thought processes. The notion that things move and my mind observes the motion carries with it the sense of *impending*, which is the root cause of fear. But time itself has nothing to do with past, present, and future. These are events in time but not time itself. Time as an experience in itself is, paradoxically, timeless.

memory

We remember something, but quite often details are unclear. John said this and I did that; or, no, John did *that* and I said *this*. Reality is a construction of what we remember. We might say that the world consists of happenings, fuzzy events, or we might call them "partially-real" events. When we attempt to determine just exactly what those events were, in effect we create them as memories. We create a past and at the same time, depending on the results of what we remember, alter ourselves by redefining our expectations of the future.

Our lives are made up of memories. Memories enable us to create rich stories, histories, excuses and explanations, all of which are put together from our reconstructed life-scenes. Our cells contain memories of our families, parents, grandparents, and even of our environment. They contain the genetic material of our ancestors, and of their ancestors who lived before human life even existed. We call these memories "racial characteristics," "genes," "DNA," and the like.

In constructing a personal memory, we follow a logical course. Our thoughts appear as pictures of causal chains of events. These events reflect our experiences or the stories we have learned from others. These experiences enter through our senses in physical processes. We usually don't remember bizarre things as really happening to us—they don't fit our version of what happened, so we simply don't remember them. We remember, instead, what we call a "classical" world, a rational world of cause and effect.

This is what consciousness does: It constructs a classical world (apparently sensible, noncontradictory, and nonparadoxical) based on a quantum world (apparently nonsensible, contradictory, and paradoxical). Thus a memory appears as a fact-like classical story and not a quantum-like superposition of paradoxical and contradictory possibilities!

We don't see what we see; we see what we remember we see. (And you can replace this phrase with "smell," "taste," "hear," "sense," and perhaps even "think.") When we see objects "out there," we not only see them, we replay all the previous information connected to them through past information "recordings."

As you move along your own story line, mind-objects—the contents of a virtual reality within the subjective realm which often appear in your dreams as dark characters—take on life and appear to you as new images, thoughts, feelings, and intuitions. You can be swept away by these images as if you were carried by a powerful wave.

The story line connects the "out there" world with the "in here" world. The wave of life moves the self from the mind-objects of the story line into the physical realm where it animates material counterparts. Then the material counterparts react and send back along these same story lines an echo wave establishing a connection between the inner virtual reality and the outer physical domain. This imaginal wave initiation/physical echo wave response results in a loop in time—where the physical activity occurs before or after the mind-object appears. When the physical activity occurs after, you experience it as wish fulfillment. When it occurs before, you see it as déjà vu or you have an inner sense of knowing what is about to happen.

The world is malleable, infinitely changeable. Not only are we capable of changing the present, but also the past.

Physicists Albert Einstein and Richard Tolman showed that if quantum mechanics describes events, then even the past is as uncertain as the future. So how do we have any past at all? The answer is that we create them! Yes. What we call the past only exists in the windmills of our mind. We in the present are responsible for our pasts, not the other way around. We are the creators of history.

Nothing is certain. No past event is in existence. Its only record exists in our neural creases.

We live in a river of time in which the source of the river (our past) and its final destination ahead of us (our future) already exist.

In what I call "the new alchemy," the future decides the present and the past falls under control of the present! In other words, the future includes the waves of unpopped possibilities and the past record of popped actualities. However, it is always possible to undo the popping and re-create the past.

a new alchemy

Everything that is, is, was, and will be. Things do not pass away in time. Every moment remains.

Think of the future as a series of pictures of yourself. Each picture portrays a different you, much as an actor changes his guise with makeup and mannerism. Each of these futures lives in the time ahead of us just as our neighbors live in the space around us. Each friendly "neighbor" calls to us, perhaps over the backyard fence or on the telephone, and invites us over. We must choose whom to visit.

Changing mind into matter requires feedback from the future, as well as what I call "feedforward" from the past. Feedback from the future appears to the mind as intuition and thought. Feedforward from the past appears as feelings and perceptions of senses.

We construct reality "out there" after the fact, with hindsight, but become aware of it as a story line, a history, even before it is completed.

We want to understand new situations as inevitable consequences of the past, and not as a result of the choices we presently make. We try to make sense of the new, basing it on the old to minimize our risk and responsibility. In other words, we make up a story to fit the sequences of our experiences. We create illusion. The maximum expression of this illusion of risk and responsibility minimization produces a feeling of security. "I failed my examination because the dog ate my term paper." "I was only following orders." "See what you made me do!" And so on.

cause and effect

If you really understand quantum physics, you understand that there is no ultimate cause and effect. Undoing the past means forgiveness and re-creating the present.

Observer effect, parallel universes, one-mind, mind creating matter—I realize that all this probably sounds wild and impossible, but I can tell you that what I used to think was unreal now seems in some ways to be more real than what I used to think was real. What used to seem real to me now seems to be more unreal!

Do you think when a new idea comes out everybody's really excited and ready to get it? No! That ain't how it works! When a new idea comes out people are very reluctant to change the way they've always thought—even when they know it's wrong. They don't want to change.

"Qwiff" is a term I coined for the quantum wave function. It's a wave that contains the potential for anything physical to appear. It's abstract and unobservable, but when it "pops," the physical world manifests.

Like the starship Enterprise on the TV show *Star Trek* whose mission it was to "seek out new worlds and boldly go where no one has gone before," the mission of the universe is to seek out all possibilities, giving all things equal opportunity to "do their thing." Our small pocket of the whole picture, our earth and its life forms, are a result of that grand plan.

But why bother? I mean, why should the universe care to follow such a plan? The answer seems to be, in order to become conscious. Consciousness necessarily demands events, real occurrences in space and time. Events can only occur if they are recorded in the mind. These events are the registrations in consciousness we call "observations."

But, such occurrences—although they themselves result in order arising out of disorder because mind has gained knowledge—require interactions that involve an overall increase in disorder. If, no matter what we do, the universe produces more and more disorder, how does any order take place?

It's here that quantum physics comes to the rescue! Order is created in the correlative behavior of qwiffs. Thus, when two objects that have previously interacted are not observed, their qwiffs become entangled and inseparable. Instead of maintaining separate qwiffs, each object joins a single qwiff. This results in a cooperative behavior between the objects even though they may no longer be in interaction with each other. This kind of cooperation plays a vital role in life and is necessary for the existence of all chemical activity.

Through such correlations, where independent objects join through mutual interaction into one qwiff, disorder is decreased. Cooperative behavior creates order, and human thought plays a major role in that enterprise. We all go where none of us have gone before.

In the opening scenes of a popular 1960s British television series that played to a select but mystified audience, the hero, known as "The Prisoner," a nameless fellow who has suddenly quit his top secret government job in England, is kidnapped from his London flat and interrogated by an antagonistic and mysterious inquisitor.

"Where am I?" The prisoner asks.

"You are here," replies the inquisitor.

"Where is 'here'?" continues the prisoner.

"Never mind that," says the inquisitor, "I am number two, you are number six."

"I am not a number," pleads the prisoner. "What do you want?"

"We want information," comes the inquisitor's reply.

After the questioning and imprisonment in a mysterious village whose location is unknown, the prisoner embarks on a number of harrowing adventures, each calculated to determine whether or not he will bend to the rules of the mysterious organization to which he has been made captive. The goal of this secret club is to force him to yield some precious knowledge, which the prisoner seems not to know, to make him a cog in the giant machinery that carries out whatever mischief the occult conglomeration chooses to perform.

Our hero resists, of course, but at the price of his own sanity.

In a sense, we are like the prisoner in the story. We live in the "information age." Facts and data seemingly jump—metaphorically and factually—quantum levels, impacting all of us. In our Internet-web-connected world, while no one would doubt the amount of information "out there," few would consider how it "secretly" affects us: information shapes our mental reality, our lives, our bodies, and the material world we inhabit.

Information transforms our everyday reality, whether we're aware of it or not. It moves and forms our thoughts and words. It makes up our vocabulary. It crosses both language and geographical barriers creating new concepts. It frightens us. It excites us. At times we feel the need to "get away from it all," meaning newspapers, the office, television, and other media. At other times we feel the need to seek out these media to see "what's happening." Information offers us new meanings to old ideas, and it affects the ways in which we conduct our relationships with others and with ourselves—even if we, like the prisoner, have never been privy to the source of that shaping and transforming intelligence.

Information (the stuff of the imaginal) not only transforms the material world, it becomes it. The adage "you are what you eat" has changed into "you are what you know" and since your knowledge ultimately depends on what information you accept as "fact," you are what you believe!

Part of why we're in such dire straits has a lot to do with our thinking. And our thinking has been influenced by desolation and death and sin and all that nonsense.

death sin desolation

In the beginning, according to what we presently know about origins, there was nothing, nothing at all. But then, something rather miraculous happened. Matter, antimatter, energy, space, time and, most significantly, information suddenly spewed into existence. It was, apparently, a kind of fluke. The universe of *nothing*, with nothing better to do, created *something*. According to our best scientific computation, if we add up all the energy in the universe, including all contained in matter and antimatter, and including the attractive kind in gravity, we come up with a big zero—it all adds up to naught. But if we add up all of the information in the universe, it appears to be nowhere near zero. Indeed it's infinite. And that is where human beings come into the picture.

What makes up things are not more things; what
makes up things are ideas, concepts, information.

The most profound creative act we experience as human beings is giving birth to our children. Great forces drive us to follow the commandment to reproduce. This drive goes far beyond the survival of humankind, indeed it goes beyond the survival of all kinds of life. It continually bubbles up in each of us; it drives our existence by producing the ideas that flow through our heads, by forming the very words we speak.

Every living thing feels this drive to create. Your dog feels it. The snake in the grass feels it. The cells of your body feel it. Bugs feel it. And plants do, too. We are all children of the vast mind-spirit that fills the universe and gave birth to it. And like our creator-spirit, each of us has the power to create. The evidence of this? We can think and speak. Here we face the miracle of the creative action—our offspring. And these offspring may take the form of our own flesh and blood or the thoughts that spring from our foreheads, as in the myth of Zeus giving birth to his daughter Athena.

To create is the master plan of the universe. Creation exacts a price for its freedom, however. It says that I must be free of the past correlations and interactions that bind me. "Untether me, let me loose of causality moorings. Do this, O Universe, and I shall work wonders for you. I will build and create machines, art forms, and people—all kinds of people and other living intelligences."

Thus it is that mind and matter cannot be truly separated. Mind is the hope of matter. The function of matter is to interact with, and thereby correlate (thus building new structures), the universe—indeed build the universe itself. The function of mind is to tear down those very structures, to analyze and decode nature's secrets, to inspect and create or re-create new structures. The universe is to be created. Mind is the creator.

matter from mind

It bears repeating: Matter arises from mind—a vast field of influence commonly envisioned as the Mind of God.

Asking a human being to explain what God is is similar to asking a fish to explain the water in which the fish swims.

When an electron scatters from a collision with a proton in a typical high-energy event, its position in space becomes dependent on its history: How fast did it move? How close did it come to a dead-center hit? Also, the proton's behavior is dependent on the collision. Although quantum mechanics does not allow us to determine the exact location of the electron or proton, it specifies exactly how the particles influence each other in a mathematical and lawful manner. So, even though we don't know where any particle is for sure, we could say that the situation is "automated."

It's as if the two particles were dancers mindlessly dancing on opposite sides of a gigantic stage, each mimicking the other's exact choreographed movements and yet each moving spontaneously. Then, with the sudden recognition that there is another dancer "there," across the stage, each dancer suddenly develops independent movements. Each has become conscious of the other and, at the same time, self-conscious. Now they must work to maintain their coordinated rhythms, the delicate nuances of each precisely copied movement.

Their dance reminds me of the ballet of Adam and Eve. While innocent and unmindful of each other, their movements in the sight of God are beautiful and symmetric. But with the inevitable bite of the apple and the serpent's way to the Tree of Knowledge, our happy, blissfully innocent couple become aware. In their awareness is a new freedom—and a new responsibility. No longer are they automatons of God. Now they are forced to work and become creators just like their Creator. Godhood disturbs the universe.

Coming into feeling—bringing mind into the body—on the one hand gives us the experience we call life, and on the other hand provides each of us with a sense of loneliness and separation from all others. We must learn to see that we are still one. We may appear to ourselves as islands, but we form a continent of life.

The illusion that mind and matter are mechanically separate may cause needless human suffering. Grasping for security will never cease as long as we think of the world as only material and fail to realize our own operational thoughts and actions within it. Perhaps the realization that mind and matter cannot be separated—that they are aspects, much like the sides of a single coin, of one and only one greater yet subtler reality—will enable the twenty-first century to be born in an atmosphere of peace not yet attainable.

natural selection

Darwin's theory of natural selection posits that creatures within a single species come into being with a random variety of characteristics, most of them good and some of them neutral or bad—"good" and "bad" referring to the ability of the creature to adapt to its environment. According to the theory, creatures with characteristics appropriate to their environment would survive, while those without characteristics appropriate to their environment would become extinct. Thus arises the notion "survival of the fittest." But, is this merely a rationalization of human behavior?

Thirty years ahead of Darwin, in 1830, scientists already knew the ideas contained in the Darwinian theory. So why weren't the ideas popular already? Because society didn't yet need them. Darwin's theory became useful, and thus popular, as the industrial age emerged. Companies were competing with fierce abandon, nations in Europe and elsewhere were seeking *lebensraum* (literally, "living room"), and kings were jockeying for advantages by playing war games with people, not pawns, on the real chessboards of life— their own countries' inhabited landscapes and cities.

In other words, Western society was rampantly exploiting the world and had begun what we now call the dog-eat-dog world view of commerce. So Darwin's *On the Origin of the Species* was just the right bible for this new enterprise. You see, science and industry go together hand in glove, not only in devising the technical means of production, but also in the employment of the alchemical forces of the imaginal realm that fuel such enterprises. Thus, particular heads of industrial states of a society, reading Darwin, create the industrial environment as "proof" that the theories are correct. These heads then adopt the Darwinian theories of science and make them popular with people ready to accept the theories and advocate them, further fueling these machines of society. This circuit becomes dominant and stable wherein what we think is supported by the environment we live in, which in turn limits and directs our thinking, thus enabling species survival.

We see an example of this every day. Companies believe that to survive they must improve their products. Within months of its introduction to the buying public, nearly any product we see on the store shelf will now bear the words "new and improved" on its label. Changing a commercial product for the better is considered a necessity for business survival, a notion that fits with Darwinian improvement of a species through natural selection. Just as you or I select the "better" product off the shelves thus insuring its continued production, Nature selects which species shall go the way of the dodo and which shall go the way of the housefly.

In other words, Darwin's theory became popular because it rationalized the human need to compete, which in turn rationalized human greed and fueled the industrial revolution.

Your knowledge of a situation changes the situation instantly. By becoming aware, you alter the outcome of the situation.

Let me tell you about a motion picture that illustrates the many facets of Niels Bohr's principle of correspondence, or as we could think of it, the "principle of illusion."

King of Hearts tells the story of a soldier who finds himself in a small French village during World War II. Not speaking French well and fearful of appearing conspicuous in Nazi-occupied France, our English soldier seeks refuge in a French insane asylum. Upon hearing of the encroaching German military machine, the "normal" townspeople desert the town, leaving it in the hands of the "crazies," who manage to find the gate of the asylum open.

The inmates soon occupy the town and take it over. In come the Germans, who are of course oblivious to all this. When they arrive on the scene it looks perfectly normal to them; after all, the French are certainly not as "normal" as the Germans who occupy them!

At this point we realize that the "crazies" are still "crazy," but because they are now "townsfolk," they appear to all intents and purposes quite normal to the occupation forces. Even later, when the Germans are driven out of town by the incoming Allied forces, the "crazies" appear "normal." It is only when the real townspeople return to their village that the secret is out and the inmates are returned to the asylum.

As I watched this film I thought that in many ways we, the people of earth, are the inmates. We're crazy and sane at the same time. We're allowed to play roles in life's great village, and as long as we continue our masquerades we remain "in character." It's this continuance, our wish for continuity, that glues us to our destinies. Underneath it all we are all quite nuts. No one alive can really take life seriously with its myriad of kaleidoscopic paradoxes; but we do.

Like us, the physical world is also a little nuts. To bend the double entendre over, each "nut" is a quantum. Bohr called his discovery "the correspondence principle" because he realized that our normal, or classical, world view is continuous. Yet his discovery of the quantum within the atom showed that atoms were fundamentally *dis*continuous in any transaction involving observers. How could the "atomic inmates" be so erratic while the "village of atoms" that make up the macroscopic universe appears so normal and orderly? Bohr's discovery showed how the "quantum insane asylum" *corresponded* with the "normal atomic village," i.e., the orderly classical world of continuous motion.

Life is a series of punctuated conscious moments. Much like the frames of a motion picture on a reel passing through a projector create an image and then vanish, our awareness of life also passes from instant to instant.

Every action we take involves this kind of on-off movement. Each time we raise an eyebrow in incredulity, or flare our nostrils in a sneer, a large number of mental events occur. As we listen to an untrustworthy politician's speech, not all of our neurons, muscle fibers, skin patches, and nerve endings want to go along to produce our incredulous sneer. Some of these bodily components, undoubtedly, want to laugh or even inhibit the actions of the other components composing the sneer. But the homeostatic majority usually wins because it not only outweighs the minority, it also can enforce its behavior in more and different ways than can our heterostatic behavior modification.

In a society of sneerers, your sneer is expected. You have learned well how to sneer. You have watched your peers sneer. You have learned just how to hold your head, to flare your nostrils, and to condescend. The society of sneerers could conceivably encompass a whole country! In such a country perhaps sneering becomes an accepted, expected norm, and if we lived in that country perhaps our normal expression would be "sneerful." Thus our faces become the face of a nation. Not only that, but our way of speaking may be shaped by our faces, our expressions literally shaping the very way we utter a word.

What would have to occur to create a shift from a sneer to a grin? Awareness and intent.

intent

"Observables" are the consequences of our actions. We "do" to observe. We must bring out or cause something to occur in order to observe anything at all.

Observation or measurement implies an observer with intelligence, a mind capable of discerning and thereby getting an impression or a perception of things. And that is what makes something go from *anything possible* to *something actual*. In other words, observation must be the creator of reality. This popularized the idea that "you create your own reality" and that quantum physics and consciousness are related. This gets spiritual when you consider who or what the ultimate observer can be.

The greatest illusion of all? The illusion of "I." This ever-present I-ness. This constant desire to turn I-ness into our "highness."

Much like Narcissus who was punished by the goddess Nemesis for resisting Echo's call, spirit embedded in matter as self—meaning body consciousness—resists spirit's call. In doing so, embodied spirit makes a primary distinction: recognizing itself as matter, it becomes entranced, lost in the image of itself as separate from spirit. An illusion, and a powerful one. Thus we, as self, begin the lifelong process of distinguishing one thing from another, a process from which we derive both joy and suffering.

feeling

Knowledge of a feeling and the feeling itself are
complementary to each other. Thus the knowledge
that you are having a feeling will alter the feeling.

Because feelings and thoughts aren't always the same, we call them "complementary." And remember, the principle of complementarity says that observation of one observable always precludes the possibility of simultaneous observation of its complement. Think about the implications . . .

Narcissus dies at the edge of the river gazing at his own reflection. Each of us suffers a similar malady as we gaze intently at the images we call our bodies. Unlike Narcissus, however, we don't just lie there, lost in our reflection. We move on, all the while feeling the loss as we miss the echo of our spirit calling to us. We live in continual stress arising from the anxiety of the ongoing battle between matter and spirit (body and soul). Some of you may object to this idea, claiming that through special techniques, meditation, spiritual practice, or simply being a good person, we may experience relief from this stress. But, like the suffering of Narcissus, the stress I refer to *must* continually arise from spirit and body opposing each other. The battle results in a continual conflict we all feel as our common human suffering. Ironically, it is this very condition that makes life worthwhile and leads to the wonderful drama of our daily reality.

Our human condition depends on the rise of spiritual stress. And here the mind enters the game. More than any other causative factor, our thoughts amplify this stress. More important than any medical care, good mental habits promote relief from this stress amplification. By good mental habits I mean simply thinking positively about every situation we encounter, even when critical thought is required.

The mere fact that there is a world at all is so miraculous, so impossible to explain, that we should, in recognition and in faith of that, be continually awestruck and continually joyful, in spite of any lacks we may feel in our daily lives. The fact that we exist in material form is no less miraculous, and it may very well be that the common suffering that we see around us and that we feel within us may be concomitant to, or the result of, the fact that we are spirits living in a material form. We are merely reflections of a single mind in a multiple-reflection looking glass.

If anyone is suffering, you are suffering, we all are suffering. To relieve such suffering, we need to recognize that each of us has created it. There is hope if we choose to hope. We are the creators of this universe. We shape its raw materials into our fantasies, which we call Reality.

suffering

The next time you find yourself suffering, try this. Become what you hate for a moment. Do this in your thoughts and discuss it with your friends and family. Turning the other cheek or loving your enemy is no act of foolish charity. It is a real solution. By accepting other's foibles as our own, we relieve our suffering.

You, in a very real sense, are holding all of us in the palm of your hand. You are the liberator of all sentient life forms.

In quantum physics we're dealing with things that jump from one place to another place seemingly without going in between. A natural question would be what's causing these jumps to occur? What's making things leap about?

The answer seems to be that observation of the things is actually causing the jumps to occur; that there's a disproportionate event that takes place which is not controllable and the mere act of observation causes things to suddenly jump into one state or another without going in between. So observation seems to be an action of consciousness because you can't observe something without being conscious of it (you need to have some idea of what you are looking for before you can observe anything). This would seem to imply that consciousness can affect matter in some way by merely observing it.

In quantum physics we have all kinds of strange twists and turns. We have objects which apparently are spatially separated yet when something happens to one of them, it instantly affects the other. Not through something traveling back and forth between the two objects, but instantaneously, as if there were no spatial separation between them at all. So in some sense, quantum non-locality is a denial of the existence of space or points to the illusory or illusionary or illusion-like quality of space. We also have similar things in time. We have things going backward and forward in time in quantum physics, which would deny the distinction between temporal events.

There is even an experiment that was done in which we couldn't distinguish when an object left a particular device, whether it started in the past or started in the present, and reached a particular goal. The actual appearance of what happened at the goal, the final point, was affected by an interference between the two possibilities of when the object actually left. So the distinction between past and present can lead to an interference, indicating that there might be no clear distinction between past and present as would seem to be obvious in our normal classical world view.

The theory of relativity tells us that time is not absolute. Fixed time intervals, say one-second units, are not equal for a moving clock and a clock at rest. As a thing moves faster and approaches the speed of light, a one-second time interval stretches, covering longer and longer time intervals as determined by the clock at rest. At the speed of light, time slows so much that it completely stops. Or, to put it another way, any time interval, no matter how short, stretches to infinite time measured by a resting clock. And so it is when we enter thought.

This is the relativity of time.

time

Becoming time is not difficult. It's like entering a flowing stream and allowing ourselves to drift along with it. As we flow with the river, the water that appeared to be rushing past when we stood upon the shore suddenly becomes quiet and motionless. It's the same when we enter the river of thought. We lose time simply because we become it.

No one knows just how the sudden popping from the imaginal possible to the real takes place. There is nothing in quantum physics itself that predicts this occurrence. Yet, this sudden "pop of reality" is the basis of Werner Heisenberg's uncertainty principle. Also called the "principle of indeterminism," the uncertainty principle reflects the inability to predict the future based on the past or based on the present. Known as the cornerstone of quantum physics, it provides an understanding of why the world appears to be made of events that cannot be connected in terms of cause and effect.

We might say that the uncertainty principle is a two-edged sword. It frees us from the past because nothing can be predetermined. It gives us the freedom to choose how we go about in the universe. But we cannot predict the exact results of our choices. We can choose, but we cannot know if our choices will turn out a certain way. The beauty is that by choosing to see spiritually, we are no longer interested in prediction. We become one with God.

*information
comes from the
yet to be*

In order to think and to express our thoughts in words, a script must appear. And it does, amazingly, allowing each of us to complete a sentence or a thought. We can express our thoughts as spoken or written words. The words just seem to pop into existence. Can we will them to appear? While will seems to play some role, it cannot push each word into existence. Somehow words come into mind. I suggest they appear because they are formed from a future's point of view. Similarly, each of us can intuit, form a hunch about what we need to do next, what's coming around the corner of our lives, so to speak. Again, I suggest that the information comes from the yet to be.

Like the Indian creator Brahma who dreams to create worlds, and the Australian aboriginal Great Spirit who dreams all of us into existence, we, too, find the source of our creative ability in our dreams. We dream in order to become aware of our future possibilities— the new ways that each of us may exist.

When we dream, the process is focused most intently on the inner world; when we're awake, it's focused most intently on the outer world. When we realize that both worlds exist, or come from the deeper more fundamental void, we're able to tap creativity and engage in creative lives.

Consciousness enables each of us to refer to ourselves as individual entities, separate from the outside world. When we're awake, once having learned to direct the stream of consciousness that bubbles within us, we're inundated with images, sensations, events, and possibilities. In normal waking consciousness we lose touch with the process and we simply take it for granted. In sleep and dreaming, without any significant interruption from outside, our bodies prepare us for direct contact.

It seems that the dream is the place where we learn how to become aware and to separate an "out there" from an "in here." The dream is a laboratory of self-creation. In this lab an entity becomes defined to itself. It's a self-referencing process, and the self-referencing process appears to be absolutely necessary for any kind of consciousness to occur. Hence we dream to awaken ourselves to the continual birthing experience of life.

dream
dream
We dream
dream
dream

to Awaken

There is evidence that over the centuries people are conditioned by their cultures to perceive things from the imaginal which are suggestive of the culture. For example, Irish culture is steeped in lore of leprechauns and fairies. It is no surprise, then, that during the eighteenth and nineteenth centuries many Irish people reported sightings of leprechauns and fairies.

In a more recent but similar way, American culture is steeped in science fiction. Over the past eighty years or so, books, movies, and television have depicted strange space creatures. From Jules Verne to *The Day the Earth Stood Still*, *Star Trek*, and *Men in Black*, the American psyche is steeped in images of aliens. So it's possible that these images originated from the imaginal realm of the authors and filmmakers, then emerged in their consciousness when they had the idea to write a book or make a movie. The images then emerged again in the dreaming brains—or altered-consciousness-state brains—of the people experiencing UFO phenomena. It may be that they're not actually seeing creatures from another space dimension emerging in our world as physical objects, but that they are tapping into the imaginal realm, which is somewhere between "real" and "fantasy," but which has some elements of both. This is not meant to belittle these experiences or to say that they are merely hallucinations. I am simply suggesting that the brain may work in more of a collective manner than we in the Western world have yet addressed.

Computers, artificial-intelligence devices, and certain robotized individuals known by their stick-to-the-rules philosophies have built-in programs to tell them what to do in novel situations. These "individuals" are intelligent only so far as the past forms the only basis for their present actions. Human beings, however, are guided by a sense of their own evolved identities in the future, which is why, in general, humans don't seem mechanical. And they're not, in any normal sense of the word.

Losing control is disturbing to our minds. We feel safest when we have control over the events of our lives. Our dream of control fits with our mechanical view of the universe. But, I have never made friends with a machine. I could imagine doing so, but I believe I would soon be disappointed. I could certainly control a machine, but after a while that would be boring. I have never been able to control a friend. Indeed, what interests me most about my friends is that they surprise me by doing the unpredictable. Similarly, the universe has many surprises in store for us through the force of disorder. Perhaps the most surprising thing about it is that this force is always with us, because we are the creators of it!

As I ponder my own existence, I see that as a species we often resist change and transformational possibilities presented to us. Perhaps we seek to simply exist on automatic pilot, not having to deal with anything new. In reviewing my own life, I see that I have often resisted transformational possibilities when they have presented themselves to me. I can't tell you that every opportunity passed over necessarily led to some personal disaster in my life, but I can tell you that when I took advantage of a possibility and turned it into a reality, it always opened a new vision of myself, my relationship with the world, and the people closest to me.

balance

Life is a balance between transformation and resistance
to transformation. Think of yourself as a dancer on the
blurry edge separating order from chaos.

The world is not as it seems and you are not as you may think you are. Quantum physics enables us to realize that the world is filled with constant change. It shows us that our observations bring the world into existence and as such provide us opportunity to change both it and ourselves.

Some people look at their lives and think something like this: "Oh God, what did I live for? Isn't it terrible that I'm going to die? Life was black when it started, bleak when I was here, and it's going to be black again when life ends! What's it all for?" In my view, this blackness and despair has been designed into God's system. We may not completely believe or even remember this design in this moment, but we have actually created all of it. The "me" that created it is not the person, the personality, that identifies itself as Fred, or Martha, or Sam—that's not that person I am speaking to or about. It's the greater essence of "I," this deeper presence, the working of consciousness itself that is in me, in you, in everyone. That I, working through this body, is the same I that is reflected in the archetypal images of Jesus, Moses, Mohammed, Krishna, all of whom reminded us, and continue to remind us, of our true essence. These beings are reflections or representations of our own identification with our greater, deeper I self.

Although I can't see God as a person or a thing, I have an experience that God is.

It's not that you have a mind and Mr. Jones has a mind and Mrs. Smith has a mind, but that you, Jones, and Smith are all of one mind. It may sound nice, may sound spiritual, to say that we're all one mind, but quantum physics actually points to its being true. The blinking on or off is a very important part of it. It indicates that mind—or the One Mind—is very much part of the physical world.

In quantum physics there is something called a "wave," an intangible, irreducible field of probability, from which all physical matter and energy arise.

The "waves" of quantum physics are ways of thinking. They're not what's going on in the physical world. Particles, particles, particles—that's real in the real world. Waves are a convenience; they're a way of thinking. Waves of possibility. Waves of probability.

When you aren't looking, it's like a wave.

When you are looking, it's like a particle.

light

Light is amazing on many counts. It is the only "thing" in the universe that is its own wave of probability as well as a physical wave carrying energy and momentum. The guiding principle offered by light is one of economy. Light gets from here to there in the most economic way possible. It follows the principle of least action.

Love can be explained in terms of the behavior of light particles—photons. All photons tend to move into the same state if given the chance; thus, in a physical sense, the phrase "light is love" is no exaggeration. Hence love represents people tending to be in a unified state of consciousness, as in, for example, lovers being of like minds. Or when we feel at one with God.

love

I discovered love at an early age. No, I don't mean the love of my family or friends, though certainly that form of love is very important. I'm talking about a larger form of love—one that connects solidly with awe, mystery, and devotion. In my early childhood, I had an interest in magic. I can't remember when I did not have this interest. It propelled me into the study of physics, because the world I perceived around me, even as a child, seemed to have a magical quality. I wondered about a lot of things. And surprisingly, anything I wondered about would eventually pop into my existence, usually not overnight, but eventually it would appear in my life. I took this quite naturally, but I didn't realize, at the time, how much this was an expression of God's love. The two are connected—love and magic.

Learning to see love and to express that love is the purpose of living this life. What's real has love at its heart; the universe is constructed from love, and that love is very much tied to our power of attention and imagination.

One would hardly suspect that scientists are motivated by love, but they are. In fact, one of the reasons science came about was to deal with that conflict that arises between the mystical feeling that, I believe, all scientists have, and the need or desire for some kind of explanation, which I attribute to the same need for security we all possess. We want to understand the world in some way. It may not be the same way in different cultures, but we want to have an understanding that enables us to cope with the various probable and improbable things that seem to happen to us from day to day, or over a lifetime. Science epitomizes the need to explain nature to ourselves and, as I see it, the need to live in the world with some kind of joy—that joy being what we call the mystical or spiritual experience.

In the story of our lives, two, often conflicting, plot lines emerge. In one story we ask ourselves how we as individuals can make the best of our lives? In the other, far more subtle, story, we deal with the recognition of ourselves as part of the human picture. In this story, we see ourselves as part of one mind, rather than as bodies in space and time, distinct and separate from others.

In this subtler story, you, as the individual, play a somewhat secondary role, although you never really lose the individual persona you've created for yourself. As you weave your life story, you find your feelings caught within the web of your own weaving. These feelings embed themselves into what I call your "dream-mask." Other masked faces show up in your dreams, and in your waking life, and are recognizable as faces of the world stage "out there." Once you're awakened to this process, you're given the opportunity to transform the face you put "out there" by changing the face you wear "in here," within the collective consciousness of images gathered throughout your life—and possibly even before you were born, but embedded deep within your genetic code.

There is no such thing as either birth or death. They both are temporary markings having to deal with the illusion that we are each a body. We can mark our bodies as having a birth and a death, but the "I" has never been born and will never die.

birth and death

A lifetime is a strange journey. It's a round trip. We end up where we began.

And finally, ponder this . . .

Being in the mystery, questioning, exploring, that's all important, but never forget that if you only get into the mysticism you miss out on what the material world is all about. Think about it. You're here and you're in a body. The fact that you've got this thing called a body—Wow! And, you've got so many things you can do with it. The world offers rich experiences to create and participate; to deny this is to be crazy. You weren't put here—or more accurately, you didn't choose to be here—in order to just get back to where you came from. Actually, you're already there. So you might as well enjoy the illusion!

More mind-bending titles by Fred Alan Wolf from Moment Point Press

Mind into Matter
Matter into Feeling
The Spiritual Universe

Visit *www.momentpoint.com*
for more information.

About the Author

You may remember him as the resident physicist on the Discovery Channel's *The Know Zone*. You may have seen him on the PBS series, *Closer to Truth*, or in the groundbreaking film *What the Bleep Do We Know!?* Respected among scientists and spiritual leaders alike for his pioneering work combining scientific and spiritual thought, Fred Alan Wolf is one of the most important pioneers in the field of consciousness.

Dr. Wolf earned his Ph.D. in theoretical physics from UCLA. He has taught at San Diego State University as well as the University of London, the University of Paris, the Hahn-Meitner Institute for Nuclear Physics in Berlin, and the Hebrew University of Jerusalem. He is also a member of the Martin Luther King, Jr. Collegium of Scholars, a National Book Award winner, and the author of twelve books.

For information regarding speaking engagements and to contact the author, please visit his website at *www.fredalanwolf.com*.